# Collins

# Edexcel GCSE 9-1
# Maths
# Foundation

*Practice Papers*

Phil Duxbury and Keith Gordon

# Contents

## SET A

## SET B

## ANSWERS

# Acknowledgements

The authors and publisher are grateful to the copyright holders for permission to use quoted materials and images.

All images are © HarperCollins*Publishers* and Shutterstock.com

Every effort has been made to trace copyright holders and obtain their permission for the use of copyright material. The author and publisher will gladly receive information enabling them to rectify any error or omission in subsequent editions. All facts are correct at time of going to press.

Published by Collins
An imprint of HarperCollins*Publishers*
1 London Bridge Street
London SE1 9GF

HarperCollins*Publishers* Macken House, 39/40 Mayor Street Upper, Dublin 1, DO1 C9W8, Ireland

© HarperCollins*Publishers* Limited 2019
ISBN 9780008321482
First published 2019
This edition published 2021
10 9 8 7 6 5 4

British Library Cataloguing in Publication Data.

A CIP record of this book is available from the British Library.

Commissioning Editor: Kerry Ferguson
Project Leader and Management: Richard Toms
Authors: Phil Duxbury and Keith Gordon
Cover Design: Sarah Duxbury and Kevin Robbins
Inside Concept Design: Ian Wrigley
Text Design and Layout: QBS Learning
Production: Karen Nulty
Printed by Ashford Colour Press Ltd

©HarperCollins*Publishers* 2019

# Collins

# Edexcel

GCSE

# Mathematics

**F**

## SET A – Paper 1 Foundation Tier (Non-Calculator)

Author: Phil Duxbury

Time allowed: 1 hour 30 minutes

**You must have:**

- Ruler graduated in centimetres and millimetres, protractor, pair of compasses, pen, HB pencil, eraser.

**You may not use a calculator**

### Instructions

- Use **black** ink or black ball-point pen.
- Answer **all** questions.
- Answer the questions in the spaces provided – *there may be more space than you need.*
- **Calculators may not be used.**
- Diagrams are NOT accurately drawn, unless otherwise indicated.
- You must **show all your working out**.

### Information

- The total mark for this paper is 80.
- The marks for **each** question are shown in brackets
  *– use this as a guide as to how much time to spend on each question.*
- Read each question carefully before you start to answer it.
- Keep an eye on the time.
- Try to answer every question.
- Check your answers if you have time at the end.

Name: _____

**1**  Calculate $3^2 \times 2^3$

**2**  Write down the first five prime numbers.

**3**  Write down the following fractions in order of size from lowest to highest.

$$\frac{1}{3}, \frac{1}{6}, \frac{4}{9}, \frac{1}{9}$$

**4**     Write 0.32 as a fraction in its simplest terms.

**5**     Change $\dfrac{7}{8}$ to a decimal.

**6**   The shape P is rotated clockwise through 90°, using a centre of rotation $(0, 1)$.

Draw the new shape on the grid provided.

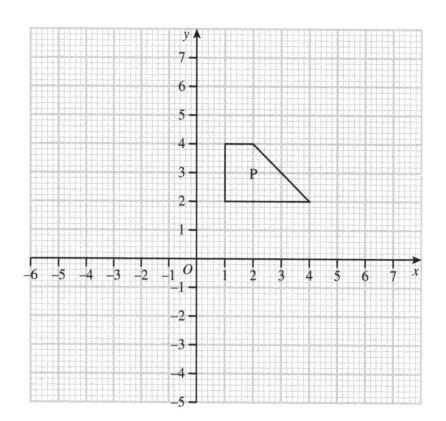

**(Total for Question 6 is 2 marks)**

**7**   Calculate $1\frac{2}{3} \times 4\frac{1}{2}$

**(Total for Question 7 is 3 marks)**

**8**    Find the area of the following semicircle, expressing your answer as a multiple of $\pi$.

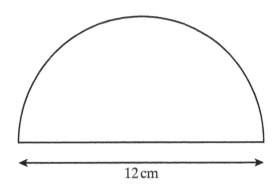

12 cm

**(Total for Question 8 is 3 marks)**

**9**    Hot dog buns come in packs of 6.

To serve enough hot dogs for everyone at his party, Gavin needs to buy enough packs to make 75 hot dogs.

Calculate the least number of packs Gavin needs to buy.

**(Total for Question 9 is 3 marks)**

**10** Work out 12% of 75.

**11** Find the lowest common multiple of 15 and 20.

**12** Solve the equation $\dfrac{x-1}{6} = \dfrac{10-x}{3}$

**13** The plan, front elevation and side elevation of a solid prism are shown below.

Plan

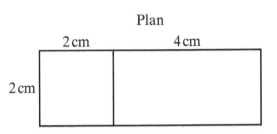

Side

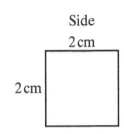

Front

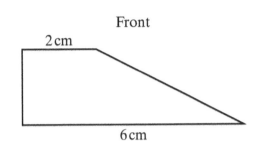

**(a)** Draw a sketch of the solid prism in three dimensions.

(1)

**(b)** Determine the volume of the prism.

(2)

**(Total for Question 13 is 3 marks)**

14      Helen can either cycle to school, take the bus or just walk.

The probability that she cycles to school on any randomly selected day is $\frac{2}{5}$ and the probability that she takes the bus is $\frac{3}{10}$

Calculate the probability that she walks to school.

**(Total for Question 14 is 3 marks)**

15      In the diagram below, calculate the size of each of the missing angles.

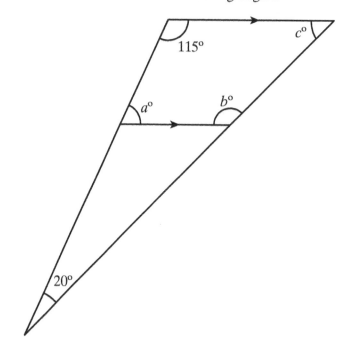

Not drawn accurately

$a =$ ........................................ °

$b =$ ........................................ °

$c =$ ........................................ °

**(Total for Question 15 is 3 marks)**

16    The following table shows the favourite sport of 60 randomly selected students.

| Sport | Frequency |
| --- | --- |
| Soccer | 16 |
| Tennis | 12 |
| Swimming | 6 |
| Athletics | 16 |
| Hockey | 10 |

(a)  Draw a pie chart to illustrate the above data.

(4)

(b)  A student is selected at random from the sample.

Find the probability that the student's favourite sport is either tennis or swimming.

(2)

**(Total for Question 16 is 6 marks)**

**17**   Jeff travels from London to Birmingham, then on to Carlisle and then on to Glasgow.

On each leg of the journey he can either travel by coach or train.

In how many different ways can Jeff travel from London to Glasgow?

**(Total for Question 17 is 2 marks)**

**18**   Work out $2.98 \times 5.1$

**(Total for Question 18 is 2 marks)**

**19**  Find the exact value for $x$ in the following triangle.

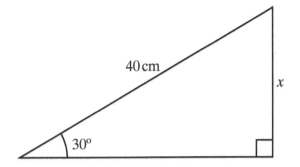

40 cm

$x$

30°

Not drawn accurately

$x =$ ........................................ cm

**(Total for Question 19 is 4 marks)**

**20**  Asif drives 200 km from London to Bath.

Assuming he travels at a constant speed of 80 km/h, calculate the time his journey takes, in hours and minutes.

**(Total for Question 20 is 3 marks)**

21   Match the correct graph with the functions below.

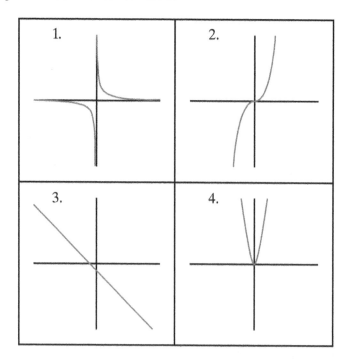

A:   $y = \dfrac{1}{x}$

(1)

B:   $y + x + 1 = 0$

(1)

C:   $y = \dfrac{1}{2}x^3$

(1)

D:   $y = 3x^2$

(1)

(Total for Question 21 is 4 marks)

22  Tim's time in the evening is spent doing either homework or relaxing, in the ratio 2 : 5

One evening, $\frac{2}{3}$ of Tim's homework is mathematics.

If Tim's evening lasts a total of 7 hours, calculate how long (in hours and minutes) Tim spends on his maths homework.

**(Total for Question 22 is 4 marks)**

23  A pair of shoes is reduced to £76 in a quick sale.

If this is a 5% reduction on the original price, find the original price of the shoes.

**(Total for Question 23 is 4 marks)**

**24** Calculate $5\frac{1}{3} \div \frac{2}{9}$

(Total for Question 24 is 4 marks)

**25** Write the following numbers in standard form.

**(a)** 33000

(1)

**(b)** 0.0082

(1)

**(c)** $0.002 \times 10^{-4}$

(1)

(Total for Question 25 is 3 marks)

**26** Expand and simplify the expression $(2x - 1)^2$

**TOTAL FOR PAPER IS 80 MARKS**

**BLANK PAGE**

# Collins

# Edexcel

## GCSE

# Mathematics

**F**

### SET A – Paper 2 Foundation Tier (Calculator)

Author: Phil Duxbury

Time allowed: 1 hour 30 minutes

**You must have:**

- Ruler graduated in centimetres and millimetres, protractor, pair of compasses, pen, HB pencil, eraser, calculator.

## Instructions

- Use **black** ink or black ball-point pen.
- Answer **all** questions.
- Answer the questions in the spaces provided – *there may be more space than you need.*
- **Calculators may be used.**
- Diagrams are NOT accurately drawn, unless otherwise indicated.
- You must **show all your working out.**

## Information

- The total mark for this paper is 80.
- The marks for **each** question are shown in brackets
  – *use this as a guide as to how much time to spend on each question.*
- Read each question carefully before you start to answer it.
- Keep an eye on the time.
- Try to answer every question.
- Check your answers if you have time at the end.

Name: _____

**1**   Write down the value of the 7 in the number 0.3071

**2**   Write down the smallest cube number greater than 100.

**3**   Write down the value of $\dfrac{3.6^4 - 102 \times 0.3}{\sqrt{5^2 + 12^2}}$ to 2 decimal places.

**4**    Consider the following list of numbers.

       12, 3, 10, 50, 5

**(a)** Find the median of this set of numbers.

(2)

**(b)** A number is now added to the list so that the median is 9.5

Find the number.

(1)

**(Total for Question 4 is 3 marks)**

**5** Solve the equation $\quad 5 - 2x = 13$

**6** Find the highest common factor (HCF) of 64 and 80.

**7** Minnie buys $p$ packets of crisps at 65p each, and $q$ packs of sandwiches at £3.00 each.

Write down an expression for the amount of money (in pence) she spends.

**8** Make $r$ the subject of this formula.

$$a = \frac{5}{20 - r}$$

**(Total for Question 8 is 4 marks)**

**9** The following table shows the heights of giraffes at a zoo.

| height ($x$ cm) | frequency |
|---|---|
| $500 \leqslant x < 510$ | 2 |
| $510 \leqslant x < 520$ | 6 |
| $520 \leqslant x < 530$ | 1 |
| $530 \leqslant x < 540$ | 4 |
| $540 \leqslant x < 550$ | 3 |

**(a)** State the modal class interval.

(1)

**(b)** Find an estimate for the mean height of the giraffes.

(2)

**(Total for Question 9 is 3 marks)**

**10**   It is suggested that the sum of two prime numbers is always an even number.

Give an example to show that this is wrong.

.................................................................................................................................................................

**11**   Find the value of $x$ in the following triangle.

Give your answer to 2 decimal places.

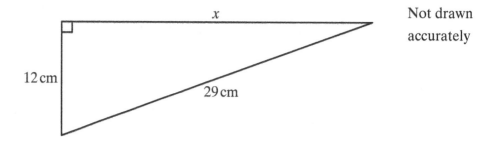

$x =$ ................................................. cm

**12**   Given vector $\mathbf{a} = \begin{pmatrix} 3 \\ -2 \end{pmatrix}$ and vector $\mathbf{b} = \begin{pmatrix} -2 \\ -1 \end{pmatrix}$, calculate the vector $2\mathbf{a} - 3\mathbf{b}$.

**13**   The following pictogram shows the holiday destinations of 210 families.

| | | | | |
|---|---|---|---|---|
| Spain | 🧍 | 🧍 | 🧍 | partial |
| France | 🧍 | 🧍 | 🧍 | |
| Germany | 🧍 | partial | | |
| Portugal | 🧍 | 🧍 | partial | |

**(a)**  The key to the pictogram is missing.

Work out what the key should be.

........................................................

(3)

**(b)**  Calculate the number of families who travelled to Spain.

........................................................

(1)

**(Total for Question 13 is 4 marks)**

14     One weekend, Katie goes shopping and spends her money on perfume and clothes in the ratio $3:7$

      If she spends £36 on perfume, calculate the amount she spends on clothes.

**(Total for Question 14 is 2 marks)**

15     The following diagram shows a series of patterns of matches.

Pattern 1        Pattern 2        Pattern 3

**(a)**   Calculate the number of matches there will be in the $4^{th}$ pattern.

(1)

**(b)**   Calculate the number of matches there will be in the $5^{th}$ pattern.

(1)

**(c)**   Find a formula for the number of matches in the $n^{th}$ pattern.

(2)

**(Total for Question 15 is 4 marks)**

**16** Solve the inequality $\dfrac{2x+7}{4} < 5$, illustrating your answer on a number line.

**(Total for Question 16 is 3 marks)**

---

**17** Here are three boxes of cereal.

| 300g | 500g | 750g |
|------|------|------|
| £1.60 | £2.60 | £3.85 |

Work out which box of cereal provides the best value for money.

**(Total for Question 17 is 3 marks)**

---

**18** Sadiq invests £1000 in a savings account paying a compound interest rate of 1.25%

For the first year only, there is a bonus 0.75% interest.

Calculate the amount (to the nearest pound) he can expect to have in his account after 5 years.

**(Total for Question 18 is 3 marks)**

**19** Kyle surveys 20 students in his school year about their favourite hobby.

He finds that 12 of them enjoy online gaming the most.

Using this result, Kyle suggests that 60% of the total students in his school would have a favourite hobby of online gaming.

Suggest **two** reasons why Kyle may be incorrect.

**(Total for Question 19 is 2 marks)**

**20**    Find the next two numbers in the following sequence.

1     4     5     9     14     23     .....     .....

**(Total for Question 20 is 2 marks)**

---

**21**    In the diagram below, angle $ABC = 140°$

Using your ruler and compasses only, construct an angle of $35°$, making your construction lines clear.

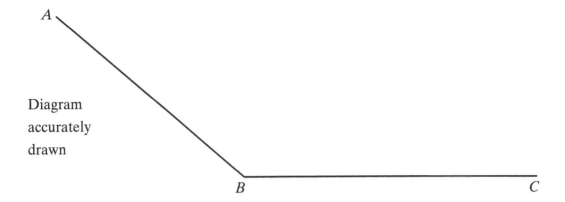

Diagram
accurately
drawn

**(Total for Question 21 is 2 marks)**

---

**22**  In the following diagram, find the size of the angle marked $x$, the size of the angle(s) marked $y$, and the size of the angle marked $z$.

In each case, give your reason(s).

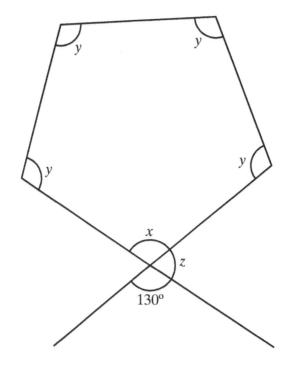

Not drawn accurately

**(a)** $x =$ ........................................................

Reason: ........................................................................

(2)

**(b)** $y =$ ........................................................

Reason: ........................................................................

(3)

**(c)** $z =$ ........................................................

Reason: ........................................................................

(2)

**(Total for Question 22 is 7 marks)**

**23**  Factorise $x^2 - 3x - 28$

........................................................................

**(Total for Question 23 is 2 marks)**

**24**

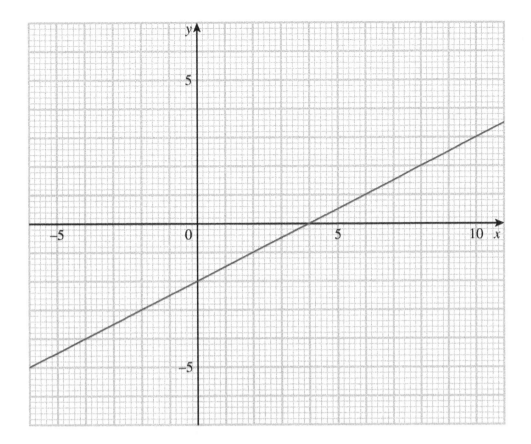

**(a)** Find the equation of the above line $L$, expressing your answer in the form $y = mx + c$

(3)

**(b)** Find the equation of the line parallel to $L$ that intersects the point $(0, 1)$.

(3)

**(Total for Question 24 is 6 marks)**

**25**    Express the ratio $16 : 25$ in the form $1 : n$, where $n$ is a decimal number.

**26**    A can of beans has radius 7.5 cm and height 11 cm.

**(a)**   The rectangular piece of paper wrapped around the tin has an 'overlap' of 2 cm.
Calculate the area of the paper (to 3 significant figures).

(4)

**(b)**   Calculate the capacity of the can (to 3 significant figures).

(2)

**(Total for Question 26 is 6 marks)**

**27**    The following diagram shows a kite.

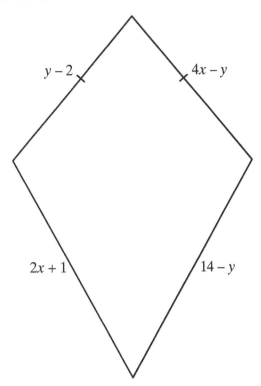

Determine the value of $x$ and the value of $y$.

$x = $ ........................................................

$y = $ ........................................................

**(Total for Question 27 is 5 marks)**

**TOTAL FOR PAPER IS 80 MARKS**

**BLANK PAGE**

# Collins

# Edexcel

GCSE

# Mathematics

**F**

### SET A – Paper 3 Foundation Tier (Calculator)

Author:  Phil Duxbury

Time allowed: 1 hour 30 minutes

**You must have:**

- Ruler graduated in centimetres and millimetres, protractor, pair of compasses, pen, HB pencil, eraser, calculator

## Instructions

- Use **black** ink or black ball-point pen.
- Answer **all** questions.
- Answer the questions in the spaces provided – *there may be more space than you need.*
- **Calculators may be used.**
- Diagrams are NOT accurately drawn, unless otherwise indicated.
- You must **show all your working out.**

## Information

- The total mark for this paper is 80.
- The marks for **each** question are shown in brackets
  – *use this as a guide as to how much time to spend on each question.*
- Read each question carefully before you start to answer it.
- Keep an eye on the time.
- Try to answer every question.
- Check your answers if you have time at the end.

Name: _____

**Answer ALL questions.**

**Write your answers in the spaces provided.**

**You must write down all the stages of your working.**

**1** Write 31 505 correct to the nearest 1000.

**(Total for Question 1 is 1 mark)**

**2** From the following list of numbers, write down the square number, and write down the prime number.

45, 46, 47, 48, 49, 50

Square number is .........................

Prime number is .........................

**(Total for Question 2 is 2 marks)**

**3** The distance between Newcastle and Carlisle is 84.65 km

Given that a road atlas has a scale of 1:625 000, calculate the distance that Newcastle and Carlisle would be apart in the atlas.

**(Total for Question 3 is 3 marks)**

**4**  The following bar chart illustrates the number of students choosing each subject at the start of their A-level courses.

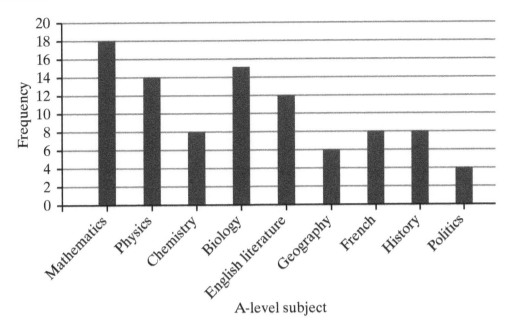

**(a)** Find the percentage of students who choose either English literature or Mathematics.

(2)

**(b)** State the modal subject choice.

(1)

**(Total for Question 4 is 3 marks)**

**5**   Tom is thinking about buying a season ticket to attend home matches at his local football club.

The price of entry to a single match is £17, while a season ticket costs £300.

**(a)**  How many matches would Tom need to attend in order for the season ticket to save him money?

(2)

**(b)**  Given that there are 21 home matches in the season, how much would Tom save if he bought a season ticket and attended all matches?

(2)

**(Total for Question 5 is 4 marks)**

**6**   The following is a scatter-diagram showing the relationship between the number of ice creams sold per day and the average temperature.

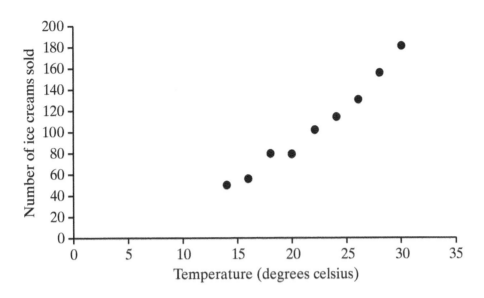

**(a)** Describe the relationship between the number of ice creams sold and the temperature.

..................................................................................................................................................................

..................................................................................................................................................................

(1)

**(b)** By drawing a line of best fit, estimate the number of ice creams that will be sold when the temperature is 8°C.

..................................................................................................................................................................

(2)

**(c)** Suggest why it may not be appropriate to draw a line of best fit for this data.

..................................................................................................................................................................

..................................................................................................................................................................

(1)

**(Total for Question 6 is 4 marks)**

**7**

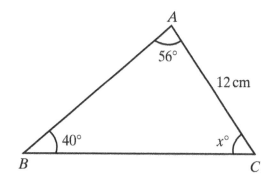

Not drawn accurately

(a) Calculate the size of the angle marked $x$.

$x =$ .................................................... °

(2)

(b) Peter says that these triangles are congruent.

Is he correct?

Give a reason for your answer.

............................................................................................................................

............................................................................................................................

(2)

**(Total for Question 7 is 4 marks)**

**8**   Consider the following number machine.

$$\text{input} \longrightarrow \boxed{-3} \longrightarrow \boxed{\div 2} \longrightarrow \text{output}$$

(a)  Find the value of the output when the input is 99.

(1)

(b)  Find the value of the input, if the output is twice its value.

(2)

**(Total for Question 8 is 3 marks)**

**9**   List the possible values of $n$ if $n$ is an integer and $-7 \leqslant n < 3$

**(Total for Question 9 is 1 mark)**

**10**  Translate the following triangle by the vector $\begin{pmatrix} 5 \\ -2 \end{pmatrix}$

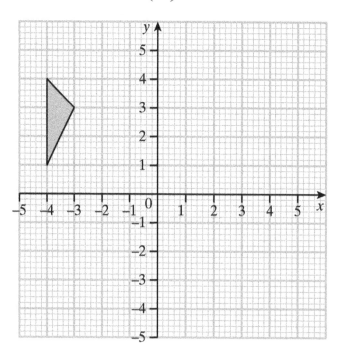

**(Total for Question 10 is 2 marks)**

---

**11**  **(a)**  Simplify   $5x - 3y + 4x + y$

(1)

**(b)**  Simplify   $(3x)^2$

(1)

**(Total for Question 11 is 2 marks)**

---

**12**  A motorcyclist travels at a steady speed of 15 m/s.

Calculate his speed in km/hr.

.......................................... km/hr

**(Total for Question 12 is 2 marks)**

---

**13**  Find the length of the side marked *x* in the following right-angled triangle.

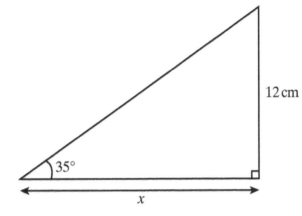

Not drawn
accurately

12 cm

35°

*x*

$x =$ .......................................... cm

**(Total for Question 13 is 2 marks)**

---

**14** State which of the following are equations, and which are identities.

Give your reasons.

**(a)** $(x - 3)^2 = x^2 + 9$

..............................................................................................................................................

(1)

**(b)** $\cos x° = \sin x°$

..............................................................................................................................................

(1)

**(c)** $x + 1 = \dfrac{x^2 - 1}{x - 1}$

..............................................................................................................................................

(1)

**(Total for Question 14 is 3 marks)**

**15** The following sector $OAB$ has a radius of 15 cm and an area of 250 cm².

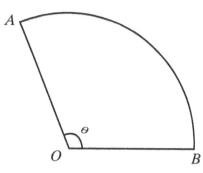

Find the size of the angle $\theta$.

$\theta = $ .............................................. °

**(Total for Question 15 is 2 marks)**

**16**    The density of gold is 19.3 g/cm³.

Calculate the mass (in kg) of a 0.1 m³ block of solid gold.

........................................................................................ kg

**(Total for Question 16 is 3 marks)**

**17**    Find the size of the angle marked $x$ in the following diagram.

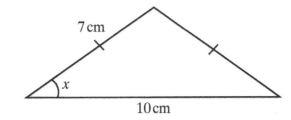

7 cm

$x$

10 cm

$x =$ ........................................................................................ °

**(Total for Question 17 is 3 marks)**

**18**    Simplify the expression $10(x - 2) - 2(x - 10)$

........................................................................................

**(Total for Question 18 is 3 marks)**

**19**   A maths test comprises of two papers: paper 1 and paper 2.

A student completes paper 1, then tackles paper 2.

The probability that a student passes paper 1 is 0.7, and the probability that a student passes paper 2 is 0.8

**(a)**   Complete the probabilities on the following tree diagram.

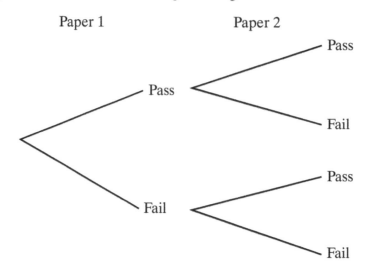

(2)

**(b)**   Find the probability that the student passes at least one of the papers.

(2)

**(Total for Question 19 is 4 marks)**

**20**   You are given the formula $\dfrac{1}{f} = \dfrac{1}{u} + \dfrac{1}{v}$

Calculate the value of $f$ when $u = 3.5$ and $v = 12.2$

**(Total for Question 20 is 2 marks)**

**21** Three apples and two bananas cost 76p.

One apple and one banana cost 32p.

Find the separate cost of each fruit.

Apple = _____ p

Banana = _____ p

**(Total for Question 21 is 4 marks)**

**22** Charlotte has a biased coin.

The probability that she throws a head is 0.6

**(a)** Charlotte throws the coin twice.

Find the probability that she throws two heads.

_____

(2)

**(b)** Charlotte throws the coin five times and finds she has thrown all heads.

What is the probability that she throws a tail on the next occasion?

_____

(1)

**(c)** Charlotte throws the coin 60 times.

How many times can she expect to throw heads?

_____

(2)

**(Total for Question 22 is 5 marks)**

**23** A triangle has vertices at $A$ (4, 3), $B$ (7, 3) and $C$ (7, 4).

**(a)** Plot the triangle on the grid below.

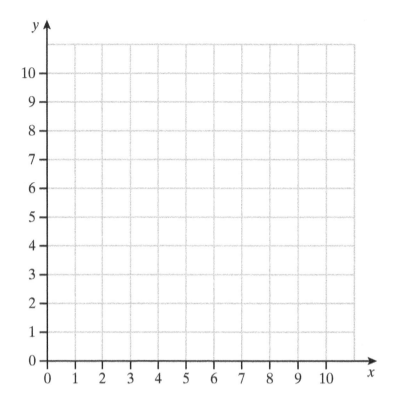

(1)

**(b)** On the grid, draw the line with equation $y = x$

(1)

**(c)** Reflect the triangle $ABC$ in the line $y = x$

(1)

**(Total for Question 23 is 3 marks)**

**24** Will pays £10 000 for a second-hand car.

Its value depreciates by 16% every year.

Find, by trial and error, how long it will take for the car's value to drop below £5000.

Show your working clearly.

**25** Simplify the expression $\dfrac{x^2 + 2x - 3}{x^2 - 9}$

**26** A line $L$ has gradient $\dfrac{4}{5}$ and passes through the point $(0, 2)$.

Find the exact coordinate where $L$ crosses the $x$-axis.

**27**  The first diagram shows part of a circle, centre $O$, radius 5 cm.

The second diagram shows a circle, centre $O$, radius $r$ cm.

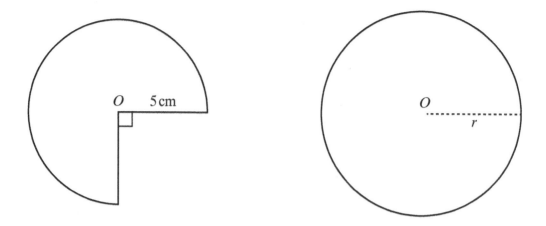

Given that both shapes have the same perimeter, calculate the value of $r$, giving your answer to 3 significant figures.

.......................................................................................................

**(Total for Question 27 is 5 marks)**

**TOTAL FOR PAPER IS 80 MARKS**

# Collins

# Edexcel

GCSE

# Mathematics

**F**

SET B – Paper 1 Foundation Tier (Non-Calculator)

Author: Keith Gordon

Time allowed: 1 hour 30 minutes

**You must have:**

- Ruler graduated in centimetres and millimetres, protractor, pair of compasses, pen, HB pencil, eraser.

**You may not use a calculator**

## Instructions

- Use **black** ink or black ball-point pen.
- Answer **all** questions.
- Answer the questions in the spaces provided – *there may be more space than you need.*
- **Calculators may not be used.**
- Diagrams are NOT accurately drawn, unless otherwise indicated.
- You must **show all your working out**.

## Information

- The total mark for this paper is 80.
- The marks for **each** question are shown in brackets
  – *use this as a guide as to how much time to spend on each question.*
- Read each question carefully before you start to answer it.
- Keep an eye on the time.
- Try to answer every question.
- Check your answers if you have time at the end.

Name: _____

**Answer ALL questions.**

**Write your answers in the spaces provided.**

**You must write down all stages in your working.**

1    How many metres are there in 3.5 kilometres?

**(Total for Question 1 is 1 mark)**

2    Here are five numbers.

    8            9            5            7            2

(a)  Work out the range of the five numbers.

(1)

(b)  Work out the median.

(1)

**(Total for Question 2 is 2 marks)**

**3**    40 people are asked to comment on the service in a restaurant.

The **pictogram** shows some of the results.

| | |
|---|---|
| Excellent | ● ● ● ● ◗ |
| Very Good | ● ● ◖ |
| Average | ● ◖ |
| Poor | ● |
| Very Poor | |

17 people said the service was excellent.

**(a)** Complete the key below.

⬤ represents _____ people

(1)

**(b)** How many people said the service was very good?

(1)

**(c)** How many people said the service was average or better?

(2)

**(d)** Complete the pictogram.                                                                    (2)

**(Total for Question 3 is 6 marks)**

**4**  **(a)**  Work out $736 + 249$

(1)

**(b)**  Work out $323 - 156$

(1)

**(c)**  Work out $6 \times 23$

(1)

**(d)**  Work out $128 \div 4$

(1)

**(Total for Question 4 is 4 marks)**

**5**  In a game a prize is hidden in one of 12 boxes.

| 1 | 2 | 3 | 4 | 5 | 6 | 7 | 8 | 9 | 10 | 11 | 12 |

Mia is playing the game.

She is told that the prize:

is not in a box that is a multiple of 3

is in a box that is a prime number

is nearer to box 1 than box 12.

Which boxes could the prize be in?

**(Total for Question 5 is 2 marks)**

**6** Mary is catching a train from Denby Dale to Manchester Airport.

She has to change trains in Huddersfield.

Here are two train timetables.

| Denby Dale | 06:24 | 07:24 | 08:24 | 09:24 | 10:24 |
|---|---|---|---|---|---|
| Huddersfield | 06:52 | 07:52 | 08:52 | 09:52 | 10:52 |

| Huddersfield | 07:02 | 08:02 | 08:35 | 09:16 | 10:02 |
|---|---|---|---|---|---|
| Manchester Airport | 07:50 | 08:50 | 09:25 | 10:05 | 10:50 |

**(a)** Mary's plane is due to take off at 12:30

She needs to be at the airport **3** hours **before** the flight is due to take off.

What is the time of the **latest** train she can catch from Denby Dale?

Circle your answer.

06:24          07:24          08:24          09:24          10:24

(1)

**(b)** Arthur is meeting someone at the airport.

He plans to get to the airport at 10:05

He catches the 08:24 from Denby Dale.

How long is his journey to the airport?

........................................

(2)

**(c)** Zak is at Huddersfield Station.

He looks at his watch.

How long will he have to wait for the next train to Manchester Airport?

........................................

(2)

**(Total for Question 6 is 5 marks)**

**7**    Eggs are delivered in trays containing 24 eggs.

A hotel orders 32 trays.

How many eggs do they order?

**(Total for Question 7 is 3 marks)**

**8**    $A(1, 2)$, $B(2, 6)$, $C(8, 6)$ and $D(7, 2)$ are the four vertices of a quadrilateral.

**(a)**  Draw the quadrilateral on the centimetre grid.

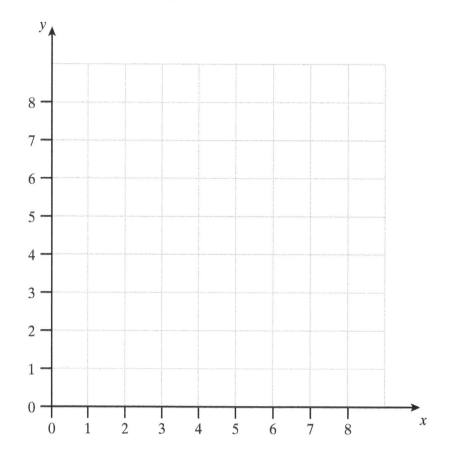

(2)

**(b)**  What type of quadrilateral is $ABCD$?

(1)

**(c)** Work out the area of *ABCD*.

(2)

**(Total for Question 8 is 5 marks)**

**9**  Show that the fraction $\frac{8}{15}$ is between $\frac{1}{3}$ and $\frac{3}{5}$

**(Total for Question 9 is 2 marks)**

**10**  **(a)**  Simplify $7a + 6a - 5a$

(1)

**(b)**  Simplify fully $2 \times 3m + 6 \times 5m$

(2)

**(Total for Question 10 is 3 marks)**

**11** The conversion graph compares acres to hectares.

Acres are a measurement of area that is commonly used in Britain.

Hectares are a metric unit of area.

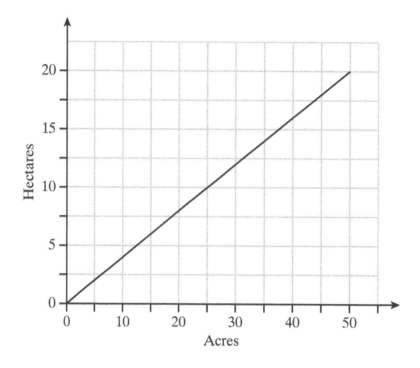

**(a)** How many acres are there in 15 hectares?

(1)

**(b)** A farm is for sale.

It has an area of 100 acres.

Farmland has an average cost of £25 000 per hectare.

Approximately how much will the farm cost?

£ ...........................................

(3)

**(Total for Question 11 is 4 marks)**

**15** Here is some information about the colour of cars in a car park.

| Colour | Frequency |
|--------|-----------|
| Blue | 7 |
| Silver | 8 |
| Red | 10 |
| White | 5 |
| Green | 6 |

Draw a fully labelled pie chart to show this information.

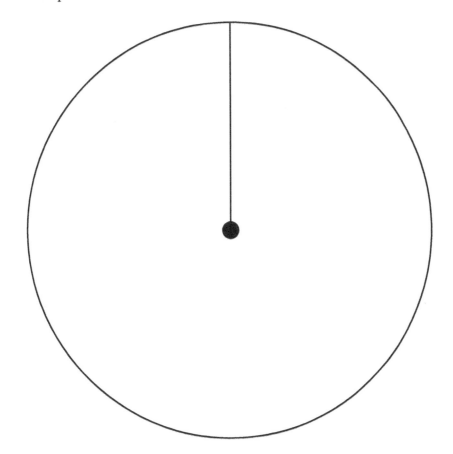

**(Total for Question 15 is 4 marks)**

**16** A cylinder has a base diameter of 20 cm and a height of 8 cm.

Calculate the volume of the cylinder.

Give your answer in terms of $\pi$.

..................................... cm$^3$

**(Total for Question 16 is 2 marks)**

**17**    Solve $3(x - 2) + 4 = \dfrac{x}{2}$

**18**    Work out the surface area of the cuboid shown.

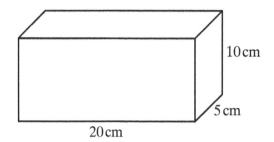

**19** Expand and simplify $4(x + 1) - 2(3x - 4)$

**(Total for Question 19 is 3 marks)**

**20** Part of a regular polygon is shown.

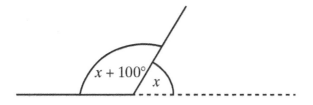

How many sides does the polygon have?

**(Total for Question 20 is 3 marks)**

**21** The graph of $y = 2x^2 - 3x - 5$ is shown.

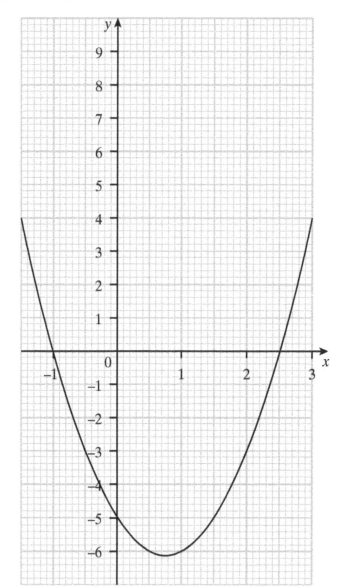

**(a)** Write down the values of $x$ when $y = 4$.

$x =$ ................................................

(2)

**(b)** Write down the coordinates of the minimum point.

................................................

(1)

**(Total for Question 21 is 3 marks)**

**22**  **(a)**  Write $2.3 \times 10^5$ as an ordinary number.

.......................................

(1)

**(b)**  Write 0.0005 in standard form.

.......................................

(1)

**(c)**  Work out $2 \times 10^4 \times 8 \times 10^3$

Give your answer in standard form.

.......................................

(2)

**(Total for Question 22 is 4 marks)**

**23**  Solve the inequality $3n + 7 > n - 4$

.......................................

**(Total for Question 23 is 3 marks)**

**24** Here is a right-angled triangle $PQR$.

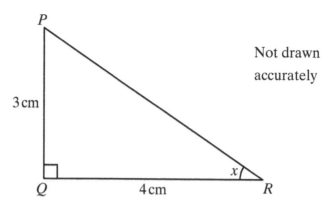

Not drawn accurately

Write down the value of the tangent of angle $x$.

$x =$ ........................................

**25** Here is a square.

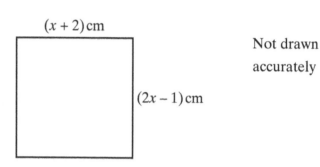

Not drawn accurately

Work out the area.

You **must** show your working.

........................................ cm$^2$

**TOTAL FOR PAPER IS 80 MARKS**

# Collins

# Edexcel

GCSE

# Mathematics

SET B – Paper 2 Foundation Tier (Calculator)

Author: Keith Gordon

**F**

Time allowed: 1 hour 30 minutes

**You must have:**

- Ruler graduated in centimetres and millimetres, protractor, pair of compasses, pen, HB pencil, eraser, calculator.

## Instructions

- Use **black** ink or black ball-point pen.
- Answer **all** questions.
- Answer the questions in the spaces provided – *there may be more space than you need.*
- **Calculators may be used.**
- Diagrams are NOT accurately drawn, unless otherwise indicated.
- You must **show all your working out**.

## Information

- The total mark for this paper is 80.
- The marks for **each** question are shown in brackets
  *– use this as a guide as to how much time to spend on each question.*
- Read each question carefully before you start to answer it.
- Keep an eye on the time.
- Try to answer every question.
- Check your answers if you have time at the end.

**Name:** _____

**Answer ALL questions.**

**Write your answers in the spaces provided.**

**You must write down all the stages of your working.**

1  (a)  Write down a number that is a multiple of **both** 5 and 8.

.......................................................................

(1)

(b)  Work out the number that is 25% **more** than 80.

.......................................................................

(1)

**(Total for Question 1 is 2 marks)**

2  Write down an expression that represents 4 less than $x$.

.......................................................................

**(Total for Question 2 is 1 mark)**

**3**     Work out the value of $3 + 4 \times 5^2$

**4**     Here are four numbered cards.

5    7    6    4

**(a)**  Write down the biggest 4-digit **odd** number that can be made with the cards.

(1)

**(b)**  How many numbers between 4000 and 5000 can be made with the four cards?

(2)

**5**    Here are six shapes.

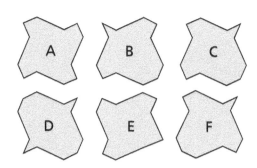

**(a)**  Which two shapes are **exactly** the same?

.........................................

(1)

**(b)**  Here is shape **A**.

Write down the order of rotational symmetry of this shape.

.........................................

(1)

**(c)**  Here is shape **D**.

Write down the number of lines of symmetry of this shape.

.........................................

(1)

**(d)**  Here is shape **F**.

What type of angle is the one marked?

.........................................

(1)

**(Total for Question 5 is 4 marks)**

**6**  A shop sells beach toys.

| Bucket | Spade | Rubber ring |
| £2.49 | £1.49 | £3.50 |

**(a)** Josie buys one of each item.

She pays with a £10 note.

How much change should she get?

£ ...........................................................

(2)

**(b)** Josie receives her change in the least number of coins possible.

Which coins did she get?

...........................................................

(2)

**(Total for Question 6 is 4 marks)**

**7**    **(a)**   Round 278 to the nearest 10.

(1)

      **(b)**   Round 3850 to the nearest 100.

(1)

**(Total for Question 7 is 2 marks)**

**8**    **(a)**   Write down the rule for continuing this sequence.

5      9      13      17      21     ....

(1)

      **(b)**   Write down the next term in the sequence in part **(a)**.

(1)

**(c)** Write down the next term in this sequence.

2    3    5    8    13    21    ....

(1)

**(d)** Write down the $n$th term of this sequence.

3    8    13    18    23    ....

(1)

**(Total for Question 8 is 4 marks)**

**9** A shape is drawn on a centimetre grid.

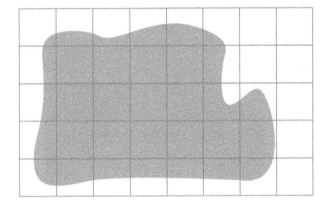

Show clearly that the area of the shape lies between 13 cm² and 33 cm².

**(Total for Question 9 is 2 marks)**

10    A running club meets on Tuesday evening.

The bar chart shows how many members came each Tuesday in January.

The information for week 4 is missing.

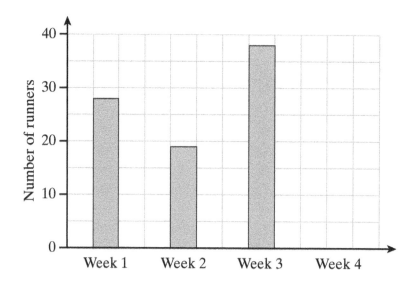

(a)  How many members came to the club in week 2?

.....................................

(1)

(b)  How many more members came in week 3 than in week 1?

.....................................

(1)

(c)  The club has 60 members.

55% of the members came to the club in week 4.

Complete the bar chart.

(3)

**(d)** The club chairman said, "On average over half of our members came each week in January".

Is she correct?

Tick a box      ☐ Yes      ☐ No      ☐ Cannot tell

Give reasons for your choice below.

(4)

**(Total for Question 10 is 9 marks)**

11     On the grid below draw:

a circle, radius 5 cm centred on A

a 6 cm by 8 cm rectangle **inside** the circle

a diagonal of the rectangle.

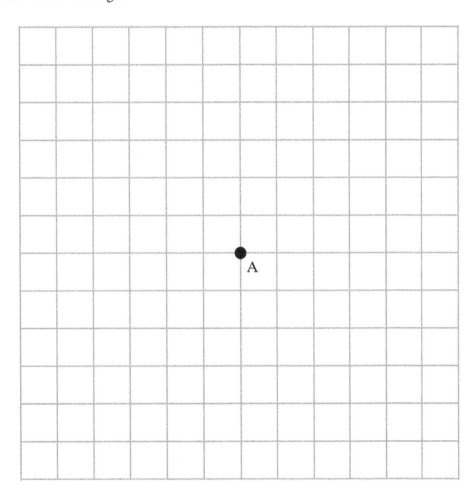

**(Total for Question 11 is 3 marks)**

**12**    **(a)**   Solve $x - 9 = 17$

(1)

     **(b)**   Solve $\dfrac{x}{4} = 8$

(1)

**(Total for Question 12 is 2 marks)**

**13**    **(a)**   Use your calculator to work out $\sqrt[3]{46.656}$

(1)

     **(b)**   Use your calculator to work out $\sqrt{105} + 19.8^2$

(1)

     **(c)**   Use estimation to show that your answer to part **(b)** is sensible.

(2)

**(Total for Question 13 is 4 marks)**

**14**  A bag contains 20 counters.

8 of the counters are yellow.

5 of the counters are blue.

The rest of the counters are red.

Work out the probability that a counter taken at random from the bag is red.

**15**  This formula is used to work out the cost of a taxi fare.

Fare = £4.00 + £2.25 for each mile + £0.75 for every minute stationary.

**(a)**  Jasmine takes a taxi that travels a distance of 7 miles and is stationary for 8 minutes.

How much was her fare?

(2)

**(b)**  Alf takes a taxi that travels a distance of 6 miles.

His fare is £21.25

For how many minutes was the taxi stationary?

(3)

**(Total for Question 15 is 5 marks)**

**16**    **(a)**   Expand $(x - 2)(x + 3)$

(2)

      **(b)**   Factorise $x^2 + 4x + 3$

(2)

**(Total for Question 16 is 4 marks)**

**17**    **(a)**   Reflect the triangle in the line $y = -1$

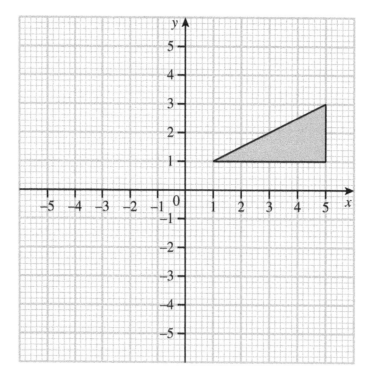

(2)

**(b)** Translate the triangle by $\begin{pmatrix} -3 \\ -4 \end{pmatrix}$

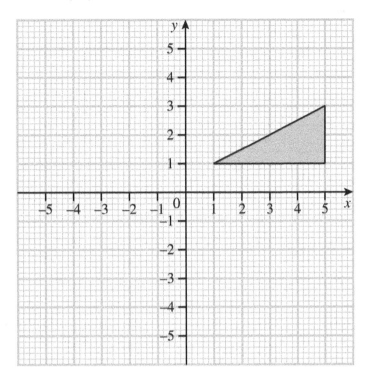

(2)

**(Total for Question 17 is 4 marks)**

---

**18** Work out the length $x$ in the triangle.

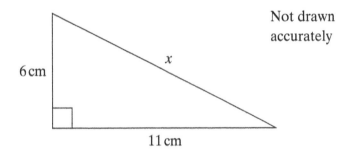

Not drawn accurately

6 cm

$x$

11 cm

$x =$ .......................................... cm

**(Total for Question 18 is 3 marks)**

---

19  The table shows the heights of some young trees.

| Height, $h$ cm | Frequency |
|---|---|
| $140 \leqslant h < 150$ | 5 |
| $150 \leqslant h < 160$ | 9 |
| $160 \leqslant h < 170$ | 12 |
| $170 \leqslant h < 180$ | 8 |
| $180 \leqslant h < 190$ | 6 |

Work out an estimate of the mean height.

**(Total for Question 19 is 3 marks)**

20  (a)  As a product of prime factors $20 = 2^2 \times 5$

Work out 28 as a product of prime factors.

(2)

(b)  Work out the least common multiple of 20 and 28.

(2)

**(Total for Question 20 is 4 marks)**

**21** Triangles *ABC* and *PQR* are similar.

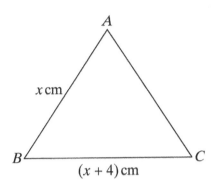

A

*x* cm

B

(*x* + 4) cm

C

P

4*x* cm

Not drawn
accurately

Q

26 cm

R

Work out the value of *x*.

$x = $ ........................................ cm

**(Total for Question 21 is 3 marks)**

**22** A washing machine is reduced by 15% in a sale.

The sale price of the washing machine is £238.

What was the original price of the washing machine?

**(Total for Question 22 is 3 marks)**

**23** Two numbers are in the ratio 2 : 5

The difference between the numbers is 36.

Work out the values of the two numbers.

**(Total for Question 23 is 3 marks)**

**24** The area of this semicircle is 201 cm² to 3 significant figures.

Not drawn
accurately

Work out the perimeter of the semicircle.

**(Total for Question 24 is 3 marks)**

**25** Using ruler and compasses only, construct an angle of 60° at $A$.

You must show your construction arcs.

$A$ ———————————————

**(Total for Question 25 is 2 marks)**

**TOTAL FOR PAPER IS 80 MARKS**

# Collins

# Edexcel

GCSE

# Mathematics

**F**

## SET B – Paper 3 Foundation Tier (Calculator)

Author:  Keith Gordon

Time allowed: 1 hour 30 minutes

**You must have:**

- Ruler graduated in centimetres and millimetres, protractor, pair of compasses, pen, HB pencil, eraser, calculator.

## Instructions

- Use **black** ink or black ball-point pen.
- Answer **all** questions.
- Answer the questions in the spaces provided – *there may be more space than you need.*
- **Calculators may be used.**
- Diagrams are NOT accurately drawn, unless otherwise indicated.
- You must **show all your working out**.

## Information

- The total mark for this paper is 80.
- The marks for **each** question are shown in brackets
  – *use this as a guide as to how much time to spend on each question.*
- Read each question carefully before you start to answer it.
- Keep an eye on the time.
- Try to answer every question.
- Check your answers if you have time at the end.

Name: ......................................................................................................................................

**Answer ALL questions.**

**Write your answers in the spaces provided.**

**You must write down all the stages of your working.**

1    Here is a shape drawn on a grid of squares.

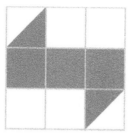

   **(a)**  Write down the number of lines of symmetry of the shape.

   ....................................................

   (1)

   **(b)**  Write down the order of rotational symmetry of the shape.

   ....................................................

   (1)

   **(Total for Question 1 is 2 marks)**

2    A piece of wood is 20 cm long measured correct to the nearest centimetre.

   Write down the interval showing the limits of the length, *l*.

   ....................................................

   **(Total for Question 2 is 2 marks)**

**3** A shape made with 10 centimetre cubes is shown on the isometric grid.

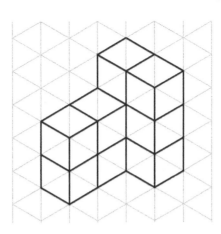

On the grids below draw the plan, side elevation and front elevation.

Plan

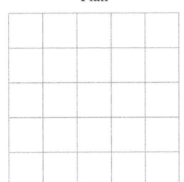

Front elevation

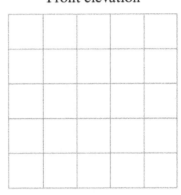

Side elevation

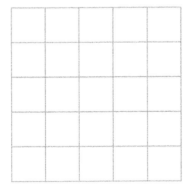

**(Total for Question 3 is 3 marks)**

**4**  Work out all the factors of 20.

.................................................................

**5**  Complete the sentence.

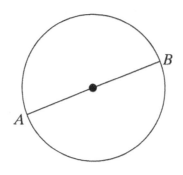

$AB$ is a ........................................................... of the circle.

**6**  Rod has 3 pairs of jeans, 4 T-shirts and 2 jackets.

He always wears jeans, a T-shirt and a jacket.

How many possible different outfits could he wear?

.................................................................

**7** Here is a number machine.

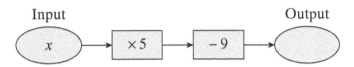

Input            Output

$x$ → ×5 → −9 →

**(a)** Work out the output when the input is 5.

...........................................

(1)

**(b)** Work out the input when the output is 11.

...........................................

(2)

**(Total for Question 7 is 3 marks)**

**8** This table shows the entry cost for using a swimming pool.

| Stacksbridge Pool<br>Open: Monday to Saturday: 6am to 8pm<br>Sunday: 8am to 2pm | | | | |
|---|---|---|---|---|
| | Adult<br>(16 and over) | Senior citizens (60 and over) | Child<br>(5 years to 15 years) | Infant<br>(Under 5) |
| **Peak**<br>Monday – Friday 6am to 9am<br>Monday – Friday 5pm to 8pm<br>Saturday and Sunday – All day | £4.50 | £4.00 | £2.50 | £1.00 |
| **Off-peak**<br>Monday – Friday 9am to 5pm | £3.50 | £3.00 | £2.00 | £1.00 |

**(a)** Neil is 18 years old and swims each day Monday to Friday from 8am to 9am.

How much does he pay each week?

(2)

**(b)** The Watson family go swimming on Friday at 1pm.

In the family are:

Mr and Mrs Watson who are both 45 years old

Andy who is 17 years old

Edie who is 13 years old

Ben who is 4 years old

Bill who is 68 years old.

How much do they pay altogether?

(3)

**(c)** The swimming pool introduces a 'Leisure Card' that costs £55 a month.

This will allow a holder of the card to swim at anytime.

Assuming that Neil would normally swim for 20 days a month, how much will he save by buying a Leisure Card?

(2)

**(Total for Question 8 is 7 marks)**

**9** Here are some coins.

Tom and Jerry divide the coins.

The amount of money they now have is in the ratio 3 : 4

What coins do they each have now?

Tom ...............................................................

Jerry ...............................................................

**(Total for Question 9 is 3 marks)**

**10** Here are eight numbers.

$$3 \qquad 8 \qquad 6 \qquad 9 \qquad 11 \qquad 12 \qquad 5 \qquad 2$$

Work out the mean of the numbers.

...............................................................

**(Total for Question 10 is 2 marks)**

**11** Work out the size of angle $x$.

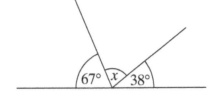

Not drawn accurately

$67°$ $x$ $38°$

$x = $ ............................................... $°$

**(Total for Question 11 is 2 marks)**

**12**  Large cuboids are 8 cm by 6 cm by 3 cm.

Small cuboids are 2 cm by 4 cm by 3 cm.

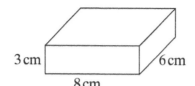 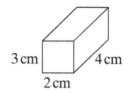

**(a)**  Show that the volume of **one** large cuboid is the same as the total volume of **six** small cuboids.

(2)

**(b)**  The large and small cuboids are stacked in alternate layers.

The bottom layer is one large cuboid.

The next layer is made from **six** small cuboids.

The total volume of the stack is 720 cm³.

How many of each type of cuboid are used in the stack?

Small cuboids ......................................

Large cuboids ......................................

(3)

**(Total for Question 12 is 5 marks)**

**13**  Cereal is sold in two sizes.

A small packet contains 350 grams and costs 79p

A large box contains 750 grams and costs £1.85

Which size is the best value?

You **must** show your working.

**(Total for Question 13 is 3 marks)**

**14** The table shows information about three journeys.

Complete the table.

| Journey | Distance | Time | Average speed |
|---------|----------|------|---------------|
| A | 32 km | | 64 km/h |
| B | | 1h 30 mins | 50 km/h |
| C | 50 km | 50 mins | |

**(Total for Question 14 is 3 marks)**

**15** A café owner records the average monthly temperature and monthly sales of ice cream over 10 months.

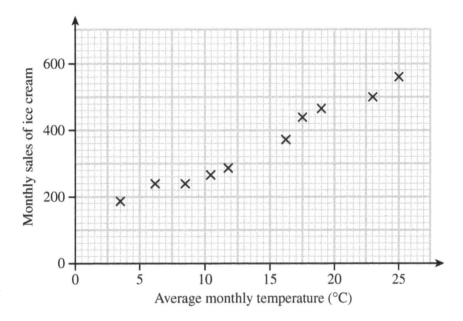

**(a)** The scatter graph shows positive correlation.

Write down the relationship between average monthly temperature and monthly sales of ice cream.

........................................................................................................................................

........................................................................................................................................

(1)

**(b)** The average monthly temperature for the next month is predicted to be 22°C.

Use the graph to estimate the sales of ice cream that month.

You **must** show your working.

(2)

**(Total for Question 15 is 3 marks)**

16    A two-digit prime number is one **more** than a square number.

Work out a possible value of the prime number.

**(Total for Question 16 is 2 marks)**

17    £3000 is invested in an account that pays 3% compound interest per year.

How much will be in the account after three years?

**(Total for Question 17 is 3 marks)**

18    **(a)** Simplify $x^3 \times x^6$

(1)

**(b)** Simplify $x^{12} \div x^2$

(1)

**(Total for Question 18 is 2 marks)**

**19**   Here are two column vectors.

$$\mathbf{a} = \begin{pmatrix} 2 \\ 3 \end{pmatrix} \quad \mathbf{b} = \begin{pmatrix} 6 \\ -2 \end{pmatrix}$$

Work out $2\mathbf{a} + \mathbf{b}$

**(Total for Question 19 is 2 marks)**

**20**   Work out the next two terms of this quadratic sequence.

3        5        8        12        17        23        ...        ...

**(Total for Question 20 is 2 marks)**

**21**   Enlarge the shape by a scale factor of $\frac{1}{3}$

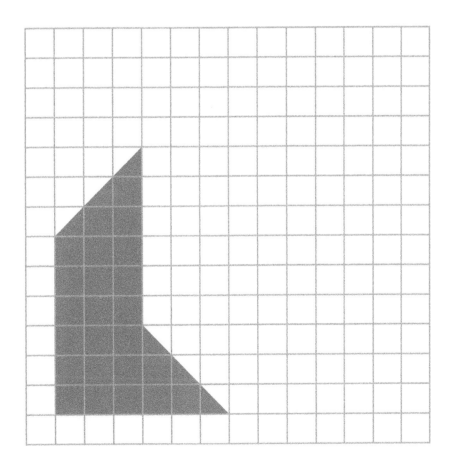

**(Total for Question 21 is 2 marks)**

**22** Two inequalities are shown.

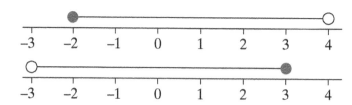

Write down the integers that are in **both** inequalities.

.....................................................

**(Total for Question 22 is 2 marks)**

**23** Here are the equations of four lines.

Line A:  $y = 3x - 4$                Line B:  $y = 4x - 3$

Line C:  $y = 3x + 3$                Line D:  $y = -4x - 4$

**(a)** Which two lines are parallel?

.....................................................

**(1)**

**(b)** Which two lines intersect on the $y$-axis?

.....................................................

**(1)**

**(Total for Question 23 is 2 marks)**

©HarperCollins*Publishers* 2019

**24** Match each graph to the equations.

Graph A      Graph B      Graph C

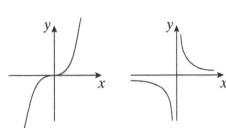

$y = x^2$ matches graph ..............................................................

$y = x^3$ matches graph ..............................................................

$y = \dfrac{1}{x}$ matches graph ..............................................................

**(Total for Question 24 is 2 marks)**

---

**25** A large candle exerts a pressure of 2 Pa on its base.

As the candle burns the pressure decreases.

After 2 hours the pressure is 0.5 Pa

Work out the rate of change of pressure.

Give your answer in Pa/hour.

**(Total for Question 25 is 2 marks)**

---

**26** Solve the simultaneous equations

$$3x + 2y = 2$$

$$x + 4y = 9$$

**(Total for Question 26 is 3 marks)**

---

**27**   A bag contains 10 balls.

4 of the balls are red and 6 are blue.

A ball is taken at random from the bag.

The ball is replaced and another ball is taken at random from the bag.

**(a)**   Complete the tree diagram.

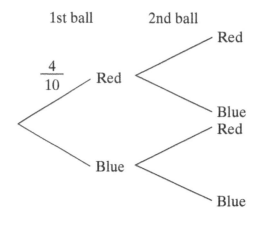

1st ball    2nd ball

$\frac{4}{10}$   Red

Red

Blue
Red

Blue

Blue

(1)

**(b)**   Use the tree diagram, or otherwise, to work out the probability that both balls were the same colour.

(3)

**(Total for Question 27 is 4 marks)**

**28**   The quadratic equation $x^2 + x - 6 = 0$ will factorise to $(x - 2)(x + 3) = 0$

Write down the solutions to the equation.

**(Total for Question 28 is 2 marks)**

**29** **(a)** Factorise $x^2 - 25$

(1)

**(b)** Show that $(x + 2)^2 - (x + 1)^2 \equiv 2x + 3$

(3)

**(Total for Question 29 is 4 marks)**

**30** **(a)** Show that the length $x$ in the triangle below is 6.36 cm to 2 decimal places.

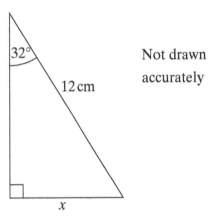

32°

12 cm

Not drawn accurately

$x$

(1)

**(b)** A cone has a half vertical angle of 32° and a slant height *l* of 12 cm.

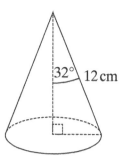

Work out the curved surface area of the cone.

The formula for the curved surface area of a cone is:

Curved surface area = π × radius of base × slant height

........................................................... cm²

(2)

**(Total for Question 30 is 3 marks)**

**TOTAL FOR PAPER IS 80 MARKS**

# Answers

**Key to abbreviations used within the answers**

M      method mark (e.g. M1 means 1 mark for method)

A      accuracy mark (e.g. A1 means 1 mark for accuracy)

B      independent marks that do not require method to be shown (e.g. B2 means 2 independent marks)

dep    dependent on previous mark

ft      follow through

oe    or equivalent

## Set A – Paper 1

| Question | Answer | Mark |
|---|---|---|
| 1 | $9 \times 8$ <br> $= 72$ | M1 <br> A1 |
| 2 | 2, 3, 5, 7, 11 | B2 |
| 3 | $\dfrac{1}{9}, \dfrac{1}{6}, \dfrac{1}{3}, \dfrac{4}{9}$ | B2 |
| 4 | $\dfrac{32}{100}$ <br><br> $= \dfrac{8}{25}$ | M1 <br><br> A1 |
| 5 | Use of valid short division method <br> $= 0.875$ | M1 <br> A1 |
| 6 | <br> Shape rotated by 90° clockwise <br> In correct place | M1 <br> A1 |
| 7 | $\dfrac{5}{3} \times \dfrac{9}{2}$ <br><br> $= \dfrac{45}{6}\left(=\dfrac{15}{2}\right)$ or $7\dfrac{1}{2}$ | M1 A1 <br><br> A1 |
| 8 | $r = 6$ <br> $\dfrac{\pi r^2}{2}$ <br> $= \dfrac{\pi \times 6^2}{2}$ <br> $= 18\pi$ cm$^2$ | B1 <br><br> M1 <br> A1 |
| 9 | $75 \div 6 = 12.5$ <br><br> So, Gavin needs 13 packs of buns | M1 A1 <br><br> A1 |
| 10 | $\dfrac{12}{100} \times 75$ or $0.12 \times 75$ <br><br> $= \dfrac{3}{25} \times 75$ <br><br> $= 9$ | M1 <br><br> A1 <br><br> A1 |
| 11 | $15 = 3 \times 5$ <br> $20 = 2 \times 2 \times 5$ <br> LCM is $2 \times 2 \times 3 \times 5 = 60$ | <br> M1 <br> M1 A1 |

| Question | Answer | Mark |
|---|---|---|
| 12 | $3(x-1) = 6(10-x)$ <br> $3x - 3 = 60 - 6x$ <br> $9x = 63$ <br> $x = 7$ | M1 A1 <br> A1 <br> A1 <br> A1 |
| 13 (a) | | B1 |
| (b) | $\dfrac{1}{2} \times (6+2) \times 2 \times 2 = 16$ cm$^3$ | M1 A1 |
| 14 | $1 - \left(\dfrac{2}{5} + \dfrac{3}{10}\right)$ <br> $1 - \left(\dfrac{4}{10} + \dfrac{3}{10}\right)$ <br> $= 1 - \dfrac{7}{10}$ <br> $= \dfrac{3}{10}$ | M1 <br><br> A1 <br><br><br> A1 |
| 15 | $a = 65°$ <br> $b = 135°$ <br> $c = 45°$ | B1 <br> B1 <br> B1 |
| 16 (a) | <br> Correct angles: $96°, 72°, 36°, 96°, 60°$ <br> Pie-chart drawn <br> Key | <br> M1 A1 <br> A1 <br> B1 |
| (b) | $\dfrac{12+6}{60}$ <br><br> $= \dfrac{18}{60}\left(= \dfrac{3}{10}\right)$ | M1 <br><br> A1 |
| 17 | $2 \times 2 \times 2$ <br> $= 8$ | M1 <br> A1 |
| 18 | Valid method of long multiplication <br> $= 15.198$ | M1 <br> A1 |
| 19 | Use of $\sin 30° = \dfrac{1}{2}$ <br> $\sin 30° = \dfrac{x}{40}$ <br> $x = 40 \sin 30°$ <br> $= 20$ cm | B1 <br><br> M1 A1 <br> A1 |

| Question | Answer | Mark |
|---|---|---|
| 20 | time = $\dfrac{\text{distance}}{\text{speed}}$ | M1 |
| | $= \dfrac{200}{80}$ | |
| | $= 2\dfrac{1}{2}$ | A1 |
| | $= 2$ hours 30 minutes | A1 |
| 21 | A: 1 | B1 |
| | B: 3 | B1 |
| | C: 2 | B1 |
| | D: 4 | B1 |
| 22 | $\dfrac{2}{7}$ of time spent on homework | B1 |
| | $\dfrac{2}{7} \times \dfrac{2}{3} \times 7$ | M1 |
| | $= \dfrac{4}{21} \times 7$ | |
| | $= \dfrac{4}{3}$ | A1 |
| | $= 1$ hour 20 minutes | A1 |
| 23 | 95% of $x = 76$ | |
| | $\dfrac{19x}{20} = 76$ | M1 A1 |
| | $x = \dfrac{76 \times 20}{19}$ | M1 |
| | $= 4 \times 20$ | |
| | $= £80$ | A1 |
| 24 | $\dfrac{16}{3} \div \dfrac{2}{9}$ | M1 |
| | $= \dfrac{16}{3} \times \dfrac{9}{2}$ | M1 A1 |
| | $= \dfrac{72}{3}$ | |
| | $= 24$ | A1 |
| 25 (a) | $3.3 \times 10^4$ | B1 |
| (b) | $8.2 \times 10^{-3}$ | B1 |
| (c) | $2 \times 10^{-7}$ | B1 |
| 26 | $(2x-1)^2 = (2x-1)(2x-1)$ | M1 |
| | $= 4x^2 - 2x - 2x + 1$ | A1 |
| | $= 4x^2 - 4x + 1$ | A1 |

**Set A – Paper 2**

| Question | Answer | Mark |
|---|---|---|
| 1 | Seven thousandths | B1 |
| 2 | 125 | B1 |
| 3 | 10.57 | B1 |
| 4 (a) | Ordering numbers: 3, 5, 10, 12, 50 | M1 |
| | Median is 10 | A1 |
| (b) | 9 | B1 |
| 5 | $-2x = 8$ | M1 |
| | $x = -4$ | A1 |
| 6 | $64 = 2 \times 2 \times 2 \times 2 \times 2 \times 2$ | M1 |
| | $80 = 2 \times 2 \times 2 \times 2 \times 5$ | |
| | HCF is $2 \times 2 \times 2 \times 2 = 16$ | A1 |
| 7 | $65p + 300q$ | B1 B1 |
| 8 | $a(20 - r) = 5$ | M1 A1 |
| | $20a - ar = 5$ or $20 - r = \dfrac{5}{a}$ | A1 |
| | $20a - 5 = ar$ | |
| | $r = \dfrac{20a - 5}{a}$ or $r = 20 - \dfrac{5}{a}$ | A1 |
| 9 (a) | $510 \leqslant x < 520$ cm | B1 |
| (b) | $\sum \dfrac{fx}{f} = \dfrac{(505 \times 2 + 515 \times 6 + 525 \times 1 + 535 \times 4 + 545 \times 3)}{16}$ | M1 |
| | $= \dfrac{8400}{16} = 525$ cm | A1 |
| 10 | 2 + any other prime | M1 |
| | e.g. $2 + 3 = 5$, so odd | A1 |
| 11 | Attempt to use Pythagoras | M1 |
| | $x^2 + 12^2 = 29^2$ | A1 |
| | $x = \sqrt{29^2 - 12^2}$ | |
| | $x = 26.40$ cm | A1 |
| 12 | $2\begin{pmatrix} 3 \\ -2 \end{pmatrix} - 3\begin{pmatrix} -2 \\ -1 \end{pmatrix} = \begin{pmatrix} 6 \\ -4 \end{pmatrix} + \begin{pmatrix} 6 \\ 3 \end{pmatrix}$ | M1 |
| | $= \begin{pmatrix} 12 \\ -1 \end{pmatrix}$ | A1 |

| Question | Answer | Mark |
|---|---|---|
| 13 (a) | 10.5 pictograms<br>One pictogram represents<br>$\frac{210}{10.5} = 20$ families | B1<br><br>M1 A1 |
| (b) | $3.5 \times 20 = 70$ families | A1 |
| 14 | $36 \div 3 = 12$<br>$= 12 \times 7$<br>$= £84$ | M1<br><br>A1 |
| 15 (a) | 13 | B1 |
| (b) | 16 | B1 |
| (c) | $3n + 1$ | M1 A1 |
| 16 | $\frac{2x + 7}{4} < 5$<br>$2x + 7 < 20$<br>$2x < 13$<br>Solution is $x < \frac{13}{2}$ or $x < 6.5$<br><br>$\frac{13}{2}$ | M1<br><br>A1<br><br>B1 |
| 17 | Any suitable method, e.g.:<br>First box: 1g costs 0.53p<br>Second box: 1g costs 0.52p<br>Third box: 1g costs 0.51p<br>So, the third box is best value<br>for money | M1<br><br><br><br>A1<br>A1 |
| 18 | $1000 \times 1.02 \times 1.0125^4$<br>$= £1072$ | M1 A1<br>A1 |
| 19 | Any two valid reasons, e.g.:<br>The sample size may be too small<br>to extrapolate<br>The sample chosen may have been<br>biased (age/gender), or otherwise<br>not representative of the school | <br><br>B1<br><br><br>B1 |
| 20 | 37<br>60 | B1<br>B1 |
| 21 | Bisect angle $ABC$ with construction<br>lines<br>Bisect angle just constructed (with<br>construction lines) | B1<br><br><br>B1 |
| 22 (a) | $x = 130°$<br>Since vertically opposite angles<br>are equal | B1<br><br>B1 |

| Question | Answer | Mark |
|---|---|---|
| (b) | Interior angles in a pentagon add<br>to 540°<br>$540° - 130° = 410°$<br>$y = \frac{410°}{4} = 102.5°$ | B1<br><br><br>M1 A1 |
| (c) | $z = 50°$<br>Since 130° and $z$ are supplementary<br>(or angles on straight line sum<br>to 180°) | B1<br><br><br>B1 |
| 23 | $(x - 7)(x + 4)$ | M1 A1 |
| 24 (a) | $m = \frac{2}{4} = \frac{1}{2}$<br>$c = -2$<br>$y = \frac{1}{2}x - 2$ | B1<br><br>B1<br><br>B1 |
| (b) | Gradient of new line is $\frac{1}{2}$<br>Equation is $y = \frac{1}{2}x + c$<br>So $y = \frac{1}{2}x + 1$ | B1<br><br>M1<br><br>A1 |
| 25 | $16 : 25 = 1 : n$<br>$16n = 25$<br>$n = 1.5625$ | M1<br><br>A1<br><br>A1 |
| 26 (a) | Circumference<br>$= 2\pi r = 2 \times \pi \times 7.5 = 47.12$ cm<br>So length of paper is '47.12' + 2 =<br>49.12 cm<br>Area $= 49.12 \times 11 = 540$ cm$^2$ | <br>M1 A1<br><br>M1<br>A1 |
| (b) | Volume $= \pi r^2 h = \pi \times 7.5^2 \times 11$<br>$= 1940$ cm$^3$ | M1<br>A1 |
| 27 | $y - 2 = 4x - y$<br>$2x + 1 = 14 - y$<br>Valid attempt to solve simultaneously<br>$x = 3$<br>$y = 7$ | M1<br>M1<br>M1<br>A1<br>A1 |

| Question | Answer | Mark |
|---|---|---|
| 1 | 32 000 | B1 |
| 2 | Square number is 49 | B1 |
| | Prime number is 47 | B1 |
| 3 | 84.65 km = 84.65 × 1000 = 8 465 000 cm | M1 |
| | $\dfrac{8\,465\,000}{625\,000} = 13.5$ cm | M1 A1 |
| 4 (a) | $\dfrac{18+12}{18+14+8+15+12+6+8+8+4} \times 100$ | M1 |
| | $= \dfrac{30}{93} \times 100$ | |
| | $= 32.3\%$ | A1 |
| (b) | Mathematics | B1 |
| 5 (a) | $\dfrac{300}{17} = 17.6\ldots$ | M1 |
| | So he needs to attend 18 matches | A1 |
| (b) | $21 \times 17 - 300$ | M1 |
| | $= £57$ | A1 |
| 6 (a) | Positive correlation (or no. of ice creams increases as temperature increases) | B1 |
| (b) | Line of best fit | B1 |
| | Approximately 15-20 ice creams | B1 |
| (c) | One good reason e.g.: relationship may not be linear / if temp. is low enough, predicted no. of ice-creams sold becomes negative | B1 |
| 7 (a) | $x = 180° - 40° - 56° = 84°$ | M1 A1 |
| (b) | No, they are not congruent | B1 |
| | since side $AC$ does not correspond to side $QR$ (ASA rule) | B1 |
| 8 (a) | 48 | B1 |
| (b) | $\dfrac{x-3}{2} = 2x$ | M1 |
| | Solve to give $x = -1$ | A1 |
| 9 | $-7, -6, -5, -4, -3, -2, -1, 0, 1, 2$ | B1 |

| Question | Answer | Mark |
|---|---|---|
| 10 | | |
| | Shape translated | B1 |
| | Correct position | B1 |
| 11 (a) | $9x - 2y$ | B1 |
| (b) | $9x^2$ | B1 |
| 12 | $15 \times \dfrac{60^2}{1000}$ | M1 |
| | $= 54$ km/hr | A1 |
| 13 | $\tan 35° = \dfrac{12}{x}$ | M1 |
| | $x = \dfrac{12}{\tan 35°} = 17.1$ cm | A1 |
| 14 (a) | Equation – only valid for certain values of $x$ | B1 |
| (b) | Equation – only valid for certain values of $x$ | B1 |
| (c) | Identity – true for all values of $x$ | B1 |
| 15 | Use $A = \dfrac{\theta}{360} \times \pi r^2$ | M1 |
| | $\theta = \dfrac{250 \times 360}{\pi \times 15^2}$ | |
| | $= 127°$ | A1 |
| 16 | 19.3 g/cm³ = (19.3/1000) / (1/1 000 000) = 19 300 kg/m³ | B1 |
| | Mass = density × volume | |
| | $= 0.1 \times 19\,300 = 1930$ kg | M1 A1 |
| 17 | Use of right angled triangle, base = 5 cm | M1 |
| | $\cos x = \dfrac{5}{7}$ | M1 |
| | $x = \cos^{-1}\left(\dfrac{5}{7}\right) = 44.4$ | A1 |

| Question | Answer | Mark |
|---|---|---|
| 18 | Valid attempt to expand brackets (at least one bracket expanded correctly) | M1 |
| | $10x - 20 - 2x + 20$ | A1 |
| | $= 8x$ | A1 |
| 19 (a) | Paper 1: 0.7, 0.3 | B1 |
| | Paper 2: 0.8, 0.2, 0.8, 0.2 | B1 |
| (b) | $1 - (0.3 \times 0.2) = 0.94$ (or $0.8 \times 0.7 + 0.8 \times 0.3 + 0.2 \times 0.7$) | M1 A1 |
| 20 | $\dfrac{1}{f} = \dfrac{1}{3.5} + \dfrac{1}{12.2} = 0.368$ | M1 |
| | $f = 2.72$ | A1 |
| 21 | $3a + 2b = 76$ $a + b = 32$ | M1 |
| | Solve simultaneously (eliminate either $a$ or $b$) | M1 |
| | $a = 12$p | A1 |
| | $b = 20$p | A1 |
| 22 (a) | $0.6 \times 0.6 = 0.36$ | M1 A1 |
| (b) | 0.4 | B1 |
| (c) | $60 \times 0.6 = 36$ | M1 A1 |
| 23 (a) | | |
| | Triangle plotted correctly | B1 |
| (b) | $y = x$ correctly drawn | B1 |
| (c) | Reflection of 'their' triangle in $y = x$ | B1 |
| 24 | Evidence of using 0.84 as a multiplier | M1 |
| | $10\,000 \times 0.84^4 = £4979$ | A1 |
| | So, 4 years | A1 |
| 25 | $\dfrac{(x+3)(x-1)}{(x+3)(x-3)}$ | M1 A1 A1 |
| | $= \dfrac{x-1}{x-3}$ | A1 |

| Question | Answer | Mark |
|---|---|---|
| 26 | Equation of $L$ is $y = \dfrac{4}{5}x + 2$ | B1 |
| | Attempt to solve $0 = \dfrac{4}{5}x + 2$ | M1 |
| | to give coordinate $\left(-\dfrac{5}{2}, 0\right)$ | A1 |
| 27 | Perimeter of shape 1 is $\dfrac{3}{4} \times 2\pi r + 10$ | M1 |
| | $= \dfrac{15\pi}{2} + 10$ | A1 |
| | Perimeter of shape 2 is $2\pi r$ | M1 |
| | $\Rightarrow 2\pi r = \dfrac{15\pi}{2} + 10$ | |
| | Setting terms equal and attempting to solve | M1 |
| | $\Rightarrow r = \dfrac{\dfrac{15\pi}{2} + 10}{2\pi} = 5.34 \text{ cm}$ | A1 |

**BLANK PAGE**

**Set B – Paper 1**

| Question | Answer | Mark | Comments |
|---|---|---|---|
| **1** | 3500 | B1 | |
| **2 (a)** | 7 | B1 | |
| **(b)** | 7 | B1 | |
| **3 (a)** | 4 | B1 | |
| **(b)** | 11 | B1 | |
| **(c)** | 4.25 + 2.75 + 1.5 or 8.5 or 17 × 2 | M1 | |
| | 34 | A1 | |
| **(d)** | 38 or 2 seen | M1 | |
| | ½ a circle drawn | A1 | |
| **4 (a)** | 985 | B1 | |
| **(b)** | 167 | B1 | |
| **(c)** | 138 | B1 | |
| **(d)** | 32 | B1 | |
| **5** | 2 and 5 | B2 | B1 for either answer and one wrong value, e.g. 2 and 7 B1 for both answer and one other value, e.g. 1, 2, 5 |
| **6 (a)** | 07:24 | B1 | |
| **(b)** | 36 + 1 + 05 | M1 | |
| | 1 h 41 m | A1 | |
| **(c)** | 09:16 seen or 16 + 20 | B1 | |
| | 36 m | B1 | |
| **7** | Clear method shown (column, box, Chinese, partition) | M1 | |
| | Correct partial calculation, e.g. 720, 48, 640, 128 or 3 out of 4 correct cells in box or Chinese methods | A1 | |
| | 768 | A1 | |

| Question | Answer | Mark | Comments |
|---|---|---|---|
| **8 (a)** | 4 correct plots | B2 | B1 for 3 correct plots or 4 plots with coordinates reversed. |
| **(b)** | Parallelogram | B1 | |
| **(c)** | 4 × 6 | M1 | |
| | 24 cm² | A1 | |
| **9** | $\frac{5}{15}$ or $\frac{9}{15}$ | M1 | |
| | $\frac{5}{15}$ and $\frac{9}{15}$ and explanation that 8 is between 5 and 9 | A1 | |
| **10 (a)** | 8a | B1 | |
| **(b)** | 6m or 30m | M1 | |
| | 36m | A1 | |
| **11 (a)** | [37, 37.5] | B1 | |
| **(b)** | 40 (hectares) | M1 | |
| | 40 × 25 000 | M1dep | |
| | £1 000 000 | A1 | |
| **12** | $\frac{4}{7}$ × 56 or $\frac{9}{11}$ × 66 | M1 | |
| | 32 or 54 | A1 | |
| | 86 | A1 | |
| **13 (a)** | Mark at $\frac{1}{3}$ | B1 | |
| **(b)** | 3 odd and 3 even numbers | B1 | e.g. 2, 3, 5, 6, 7, 8 is B2 2, 3, 4, 5, 6, 8 is B1 2, 3, 4, 5, 7, 8 is B1 2, 3, 5, 6, 7, 9 is B0 |
| | 2 multiples of 3 | B1 | |
| **14** | ABC or ACB = 80 | M1 | |
| | ACD = 100 | M1dep | |
| | 40° | A1 | |

| Question | Answer | Mark | Comments |
|---|---|---|---|
| 15 | $360 \div 36 = 10$ | M1 | |
| | Angles calculated as 70, 80, 100, 50 and 60 | M1dep | |
| | Angles accurately drawn | A1 | |
| | Sectors labelled | A1 | |
| 16 | $\pi \times 10^2 \times 8$ | M1 | |
| | $800\pi$ cm$^3$ | A1 | |
| 17 | $6x - 12 + 8 = x$ | M1 | |
| | $5x = 4$ | M1dep | |
| | $x = 0.8$ oe | A1 | |
| 18 | Area any face, i.e. $20 \times 5$ or $100$ etc. | M1 | |
| | $2 \times 100 + 2 \times 50 + 2 \times 200$ | M1dep | |
| | $700$ cm$^2$ | A1 | |
| 19 | $4x + 4 - 6x + 8$ | M1 | M1 for 3 terms correct |
| | $4x + 4 - 6x + 8$ | A1 | A1 for 4 terms correct |
| | $-2x + 12$ | A1ft | ft on M1, e.g. $4x + 1 - 6x - 8 = -2x - 7$ is M1, A0, A1ft |
| 20 | $2x + 100 = 180$ | M1 | |
| | $360 \div 40$ | M1dep | |
| | $9$ | A1 | |
| 21 (a) | $-1.5$ and $3$ | B2 | B1 each answer |
| (b) | $(0.75, -6.1)$ | B1 | |
| 22 (a) | $230\,000$ | B1 | |
| (b) | $5 \times 10^{-4}$ | B1 | |
| (c) | $1.6 \times 10^8$ | B2 | B1 for $16 \times 10^7$ |
| 23 | $2n > -11$ | M2 | M1 for $2n > 3$ or $2n > -3$ or $4n > -11$ |
| | $n > -5.5$ | A1ft | ft on M1, e.g. $n > 1.5$ |

| Question | Answer | Mark | Comments |
|---|---|---|---|
| 24 | $\dfrac{3}{4}$ | B1 | |
| 25 | $x + 2 = 2x - 1$ | M1 | |
| | $x = 3$ | A1 | |
| | $3 + 2$ or $2 \times 3 - 1$ | M1dep | |
| | $5$ | A1 | |
| | $25$ | A1 | |

**Set B – Paper 2**

| Question | Answer | Mark | Comments |
|---|---|---|---|
| 1 (a) | Any multiple of 40, e.g. 40 | B1 | |
| (b) | 100 | B1 | |
| 2 | $x - 4$ | B1 | |
| 3 | 103 | B1 | |
| 4 (a) | 7645 | B1 | |
| (b) | Any 2 numbers shown, e.g. 4675, 4657 etc. | M1 | |
| | 6 | A1 | |
| 5 (a) | B and F | B1 | |
| (b) | 4 | B1 | |
| (c) | 2 | B1 | |
| (d) | Reflex | B1 | |
| 6 (a) | 7.48 or 748 seen | M1 | |
| | £2.52 | A1 | |
| (b) | £2, 50p, 2p | B2ft | ft least number of coins for their answer for (a) B1 for any correct combination of coins but not least number |
| 7 (a) | 280 | B1 | |
| (b) | 3900 | B1 | |
| 8 (a) | Add 4 each time | B1 | |
| (b) | 25 | B1ft | ft their rule |
| (c) | 34 | B1 | |
| (d) | $5n - 2$ | B1 | |

| Question | Answer | Mark | Comments |
|---|---|---|---|
| 9 | Marks on diagram showing counting of 13 whole squares within or 33 outside shape | M1 | |
| | Explanation that area must be between these limits | A1 | |
| 10 (a) | 19 | B1 | |
| (b) | 10 | B1 | |
| (c) | 0.55 × 60 oe | M1 | |
| | 33 | A1 | |
| | Bar drawn to 33 | A1 | |
| (d) | 28 + 19 + 38 + their week 4 or 118 | M1 | |
| | 240 seen | B1 | |
| | 0.5 × 240 or 120 | M1 | |
| | Correct conclusion based on their total (No if correct) | A1 | |
| 11 | | B3 | B1 for circle B1 for rectangle (may be a different orientation) B1 for either diagonal (allow both drawn) |
| 12 (a) | 26 | B1 | |
| (b) | 32 | B1 | |
| 13 (a) | 3.6 | B1 | |
| (b) | 402.(2…) | B1 | |
| (c) | Either value rounded to 1 sf e.g. 100 or 20 | M1 | |
| | 10 + 400 = 410 | A1 | |
| 14 | $\dfrac{7}{20}$ | B2 | B1 for 7 seen |
| 15 (a) | 4 + 7 × 2.25 + 8 × 0.75 | M1 | Allow mixed units |
| | 25.75 | A1 | |

| Question | Answer | Mark | Comments |
|---|---|---|---|
| (b) | 21.25 – 6 × 2.25 – 4 or 3.75 | M1 | Allow mixed units |
| | Their 3.75 ÷ 0.75 | M1dep | |
| | 5 | A1 | |
| 16 (a) | $x^2 - 2x + 3x - 6$ | M1 | 4 terms, with one in $x^2$, 2 in $x$ and a constant term |
| | $x^2 + x - 6$ | A1 | |
| (b) | $(x + a)(x + b)$ where $ab = \pm3$ | M1 | |
| | $(x + 1)(x + 3)$ | A1 | |
| 17 (a) | Correct reflection, i.e. (5, –3), (1, –3), (5, –5) | B2 | B1 for reflection in $x = -1$ |
| (b) | Correct translation, i.e. (–2, –3), (2, –3), (2, –1) | B2 | B1 for correct translation of one vector component |
| 18 | $6^2 + 11^2$ | M1 | |
| | $\sqrt{157}$ | M1dep | |
| | 12.5… | A1 | |
| 19 | 5 × 145 + 9 × 155 + 12 × 165 + 8 × 175 + 6 × 185 or 6610 | M1 | |
| | 6610 ÷ 40 | M1dep | |
| | 165.25 | A1 | |
| 20 (a) | Any product including a prime that makes 28 | M1 | |
| | 2 × 2 × 7 or $2^2$ × 7 | A1 | |
| (b) | 2 × 2 × 5 × 7 | M1 | |
| | 140 | A1 | |
| 21 | $4(x + 4) = 26$ | M1 | |
| | $4x = 10$ | M1dep | |
| | 2.5 | A1 | |
| 22 | 0.85 | B1 | |
| | 238 ÷ 0.85 | M1 | |
| | 280 | A1 | |

| Question | Answer | Mark | Comments |
|---|---|---|---|
| 23 | $36 \div 3$ or 12 | M1 | |
| | $2 \times 12$ or $5 \times 12$ | M1dep | |
| | 24 and 60 | A1 | |
| 24 | $\sqrt{\dfrac{402}{\pi}}$ or 11.3… | M1 | |
| | $11.3 \times \pi + 2 \times 11.3$ | M1dep | |
| | [58, 58.2] | A1 | |
| 25 | Arc from A cutting given ray | M1 | |
| | Arc centred on intersection and crossing original arc plus line drawn. | A1 | Angle must be between [58, 62] |

## Set B – Paper 3

| Question | Answer | Mark | Comments |
|---|---|---|---|
| 1 (a) | 0 | B1 | |
| (b) | 2 | B1 | |
| 2 | 19.5 or 20.5 seen | B1 | |
| | $19.5 \leqslant l < 20.5$ | B1 | |
| 3 | Plan<br><br><br><br>Front Elevation<br><br><br><br>Side Elevation<br><br> | B3 | B1 each<br>Accept front and side elevation labelled the other way round |
| 4 | 1, 2, 4, 5 10, 20 | B2 | B1 for 4 or 5 factors |
| 5 | Diameter | B1 | |

| Question | Answer | Mark | Comments |
|---|---|---|---|
| 6 | $3 \times 4 \times 2$ | M1 | |
| | 24 | A1 | |
| 7 (a) | 16 | B1 | |
| (b) | 4 | B2 | B1 for 100<br>B1 for 0.4 |
| 8 (a) | $5 \times 4.50$ | M1 | |
| | £22.50 | A1 | |
| (b) | $3 \times 3.50 + 3.00 + 2.00 + 1.00$ | M1 | M1 for 5 people identified and off peak prices |
| | $3 \times 3.50 + 3.00 + 2.00 + 1.00$ | A1 | All six identified and off peak prices |
| | £16.50 | A1 | 16.5 is A0<br>SC2 for 21 |
| (c) | $20 \times 4.50 - 55$ | M1 | |
| | £35 | A1 | |
| 9 | $98 \div 7$ or 14 | M1 | |
| | 42 or 56 | A1 | |
| | Tom 20, 10, 10, 2<br>Jerry 50, 5, 1 | A1 | Either order |
| 10 | $56 \div 8$ | M1 | |
| | 7 | A1 | |
| 11 | $180 - 67 - 38$ | M1 | |
| | $75°$ | A1 | |
| 12 (a) | $3 \times 8 \times 6$ or 144<br>or $3 \times 2 \times 4$ or 24 | M1 | |
| | $144 \div 24$ (= 6) | A1 | |
| (b) | $720 \div 144$ or 5 (layers) | M1 | |
| | Small 12 | A1 | |
| | Large 3 | A1 | |
| 13 | $350 \div 79$ or $750 \div 185$ | M1 | Allow mixed units |
| | 4.43… or 4.05… | A1 | |
| | small packet | A1 | |

| Question | Answer | Mark | Comments |
|---|---|---|---|
| 14 | 30 mins or 0.5 hours | B1 | |
| | 75 km | B1 | |
| | 60 km/h | B1 | |
| 15 (a) | More ice cream sold as temperature increases | B1 | |
| (b) | Line of best fit | M1 | |
| | 480 | A1ft | ft their line of best fit |
| 16 | 17 or 37 | B2 | B1 for 26, 50, 65 or 82 |
| 17 | 1.03 | B1 | |
| | $3000 \times 1.03^3$ | M1 | |
| | 3278.18 | A1 | |
| 18 (a) | $x^9$ | B1 | |
| (b) | $x^{10}$ | B1 | |
| 19 | $\begin{pmatrix} 10 \\ 4 \end{pmatrix}$ | B2 | B1 for each component |
| 20 | 30 | B1 | |
| | 38 | B1 | |
| 21 | | B2 | B1 for any enlargement that reduces the size of the shape and keeps the sides in relative ratio. B1 for any 3 sides correct. |
| 22 | –2, –1, 0, 1, 2, 3 | B2 | B1 for –3, –2, –1, 0, 1, 2, 3 B1 for –2, –1, 0, 1, 2, 3, 4 |
| 23 (a) | A and C | B1 | |
| (b) | A and D | B1 | |
| 24 | C A B | B2 | B1 for 1 correct |
| 25 | $1.5 \div 2$ | M1 | |
| | 0.75 | A1 | |

| Question | Answer | Mark | Comments |
|---|---|---|---|
| 26 | $3x + 2y = 2$ and $3x + 12y = 27$ or $6x + 4y = 4$ and $x + 4y = 9$ | M1 | |
| | $x = -1$ | A1 | |
| | $y = 2.5$ | A1 | |
| 27 (a) | $\frac{4}{10}$ marked on red and $\frac{6}{10}$ marked on blue | B1 | |
| (b) | $\frac{4}{10} \times \frac{4}{10}$ or $\frac{6}{10} \times \frac{6}{10}$ | M1 | |
| | $\frac{4}{10} \times \frac{4}{10} +$ $\frac{6}{10} \times \frac{6}{10}$ | M1dep | |
| | 0.52 | A1 | oe |
| 28 | 2 | B1 | |
| | –3 | B1 | |
| 29 (a) | $(x + 5)(x - 5)$ | B1 | |
| (b) | $x^2 + 4x + 4$ or $x^2 + 2x + 1$ | M1 | $(x + 2 + x + 1)$ $(x + 2 - (x + 1))$ |
| | $x^2 + 4x + 4 - (x^2 + 2x + 1)$ | M1dep | $(2x + 3)(1)$ |
| | Shows subtraction of terms clearly | A1 | |
| 30 (a) | $12 \times \sin 32 = 6.359…$ | B1 | |
| (b) | $\pi \times 6.36 \times 12$ | M1 | |
| | [236.6, 240] | A1 | |

**BLANK PAGE**

**BLANK PAGE**